Building Blocks of Biology

Genes and Genetics

Written by Alex Woolf

Illustrated by Ruth Bennett

World Book

www.worldbook.com

Co-published by agreement between Shi Tu Hui and World Book, Inc.

Shi Tu Hui
Room 1807, Block 1,
#3 West Dawang Road
Chaoyang District, Beijing 100025
P.R. China

World Book, Inc.
180 North LaSalle Street
Suite 900
Chicago, Illinois 60601
USA

Library of Congress Control Number: 2025942742

Building Blocks of Biology
ISBN: 978-0-7166-6737-7 (set, hard cover)

Genes and Genetics
ISBN: 978-0-7166-6745-2 (hard cover)

Also available as:
ISBN: 978-0-7166-6765-0 (e-book)
ISBN: 978-0-7166-6755-1 (soft cover)

WORLD BOOK STAFF

Editorial

Vice President
Tom Evans

Senior Manager, New Content
Jeff De La Rosa

Proofreader
Nathalie Strassheim

Graphics and Design

Senior Visual Communications Designer
Melanie Bender

Acknowledgments
Writer: Alex Woolf
Illustrator: Ruth Bennett/The Bright Agency

TABLE OF CONTENTS

There is a glossary on page 40. Terms defined in the glossary are in type **that looks like this** on their first appearance.

BABY PHOTOS
One afternoon...
I love looking at old photos!
There's me when I was a cub!

This is me as a sprout...
Wow! You look so little!

This is me as a fry–that's a baby fish...
BEEP
BOOP
And these are my parents!

Have you ever noticed that we grow up to look much like our parents?
You got that right!
Why is that?

TRAITS
It's because our parents pass on certain **traits** to us.
A trait is a particular characteristic of a living thing.
See these three leaves growing from the top of my head?
That's a trait I inherited from my parents!
I have brown and white patches on my fur, a trait I share with my parents!

What three examples of traits do Root, Fur, and Fin mention?
See page 40 for answers.

Later, in Root's laboratory...
We inherit many traits from our parents...
We carry information about these traits in the cells of our bodies.
Hmm...
Hey, what are you looking at?
It looks like a flake of human skin...
That means a human has been sneaking around your laboratory, Root!

I wonder who it was...
You said information about our traits is inside our cells...
Maybe we could identify the intruder from the cells in this skin!
Brilliant idea, Fin! First, we'll need to make ourselves very small...
HUMMMMMM
WHIRRRRR
Time for another journey aboard my Nanoship!

CHROMOSOMES AND DNA

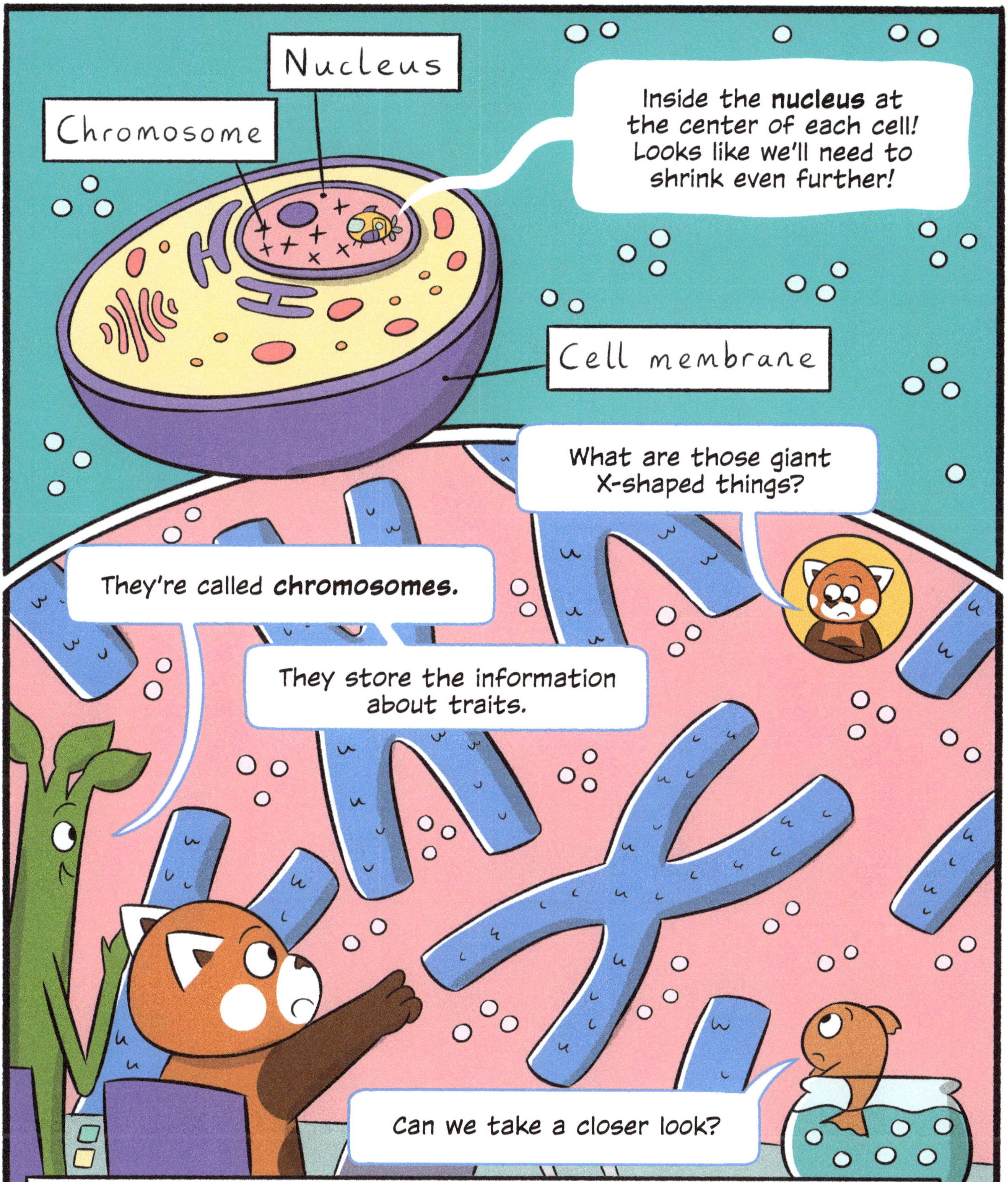

Fill in the blanks. Information on traits is stored in X-shaped structures called _____ . These are found in the _____ of the cell.
See page 40 for answers.

When you said the chromosome contains information, I expected it to look like a giant book...
This chromosome contains more information than the biggest book you've ever seen!
But what's it made of?
Chromosomes are tiny, threadlike structures inside the cells of every living thing. Every organism has a certain number of chromosomes in its cells. In most plants and animals, chromosomes come in pairs. Human cells have 23 pairs, for a total of 46 chromosomes (23 x 2 = 46).
SHR-I-N-K
Let's shrink down further for a closer look!
Wow! It looks like a twisting ladder.

This is what a chromosome is made of... DNA!
Its "twisting ladder" shape is called a *double helix.*
DNA is a complex *molecule* (group of atoms bound together). *DNA* stands for *deoxyribonucleic acid.*
A
T
Adenine
Thymine
Wow! The rungs are all different!
Base pairs
G
C
Guanine
Cytosine
Is that where the information is?
Yes, those rungs are made up of four different chemicals, called bases. The bases are adenine (A), guanine (G), cytosine (C), and thymine (T).
Bases pair up to form base pairs.
The order, or sequence, of base pairs is like a code that helps determine our traits.

GENES AND ALLELES
So, our DNA is kind of like a recipe book for our bodies?
But a recipe book tells somebody how to make a cake or a soup or something...
It's quite similar with DNA... The sequence of the bases tells our cells how to make our bodies.
A DNA sequence that contains the instruction for a particular trait is called a **gene**...
GENE
DNA

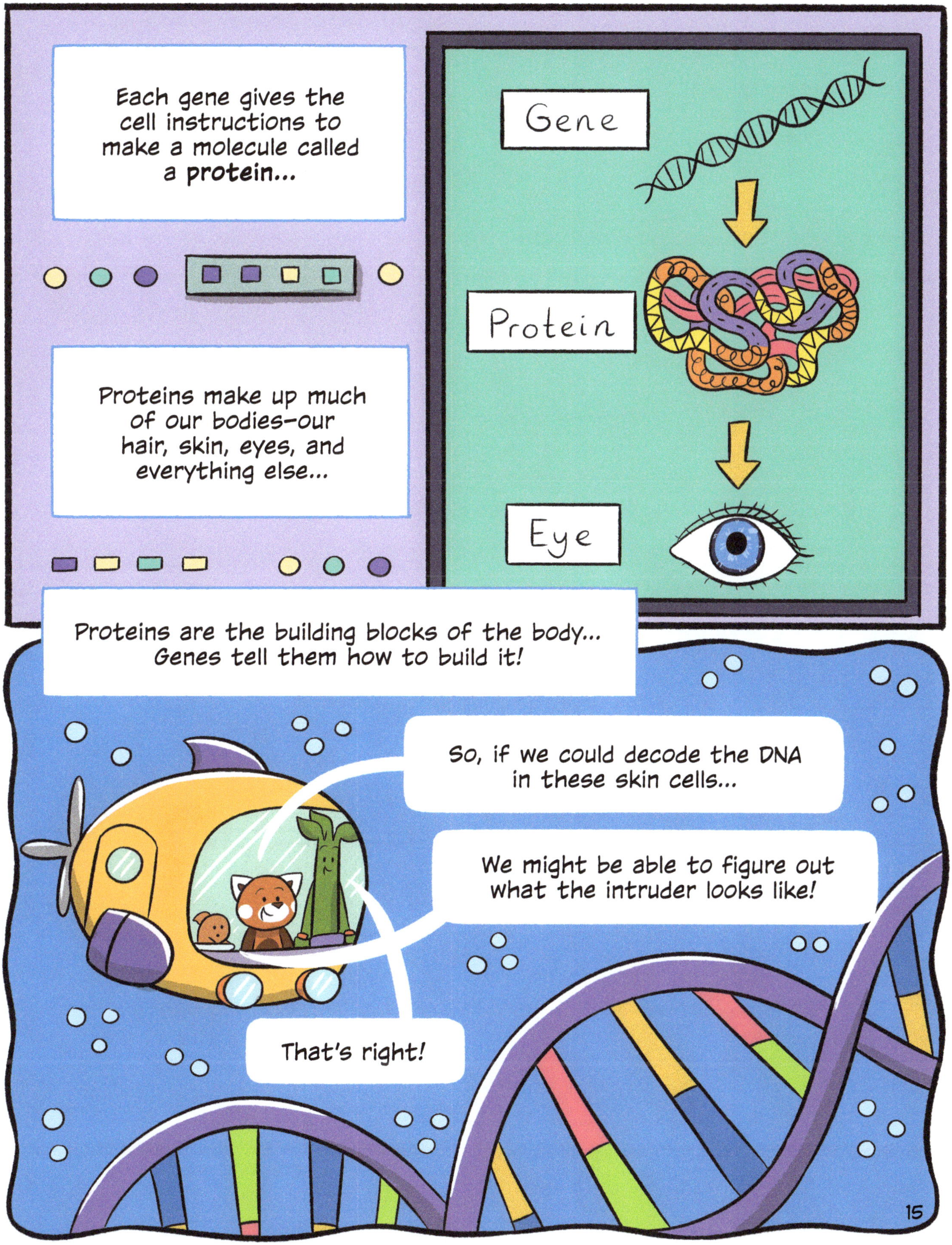

Each gene gives the cell instructions to make a molecule called a **protein**...
Gene
Protein
Eye
Proteins make up much of our bodies—our hair, skin, eyes, and everything else...
Proteins are the building blocks of the body... Genes tell them how to build it!
So, if we could decode the DNA in these skin cells...
We might be able to figure out what the intruder looks like!
That's right!

Most genes are the same from person to person... That's why people look pretty much alike!
But, a few genes can vary from one person to the next... The variant genes are called **alleles.**
Alleles are responsible for traits that vary from person to person, for example skin tone or hair color.
Fur, both your parents passed on to you the allele for brown eyes.
What if Fur had gotten a brown eye allele from one parent and a blue eye allele from the other?
Would he have one brown eye and one blue eye?

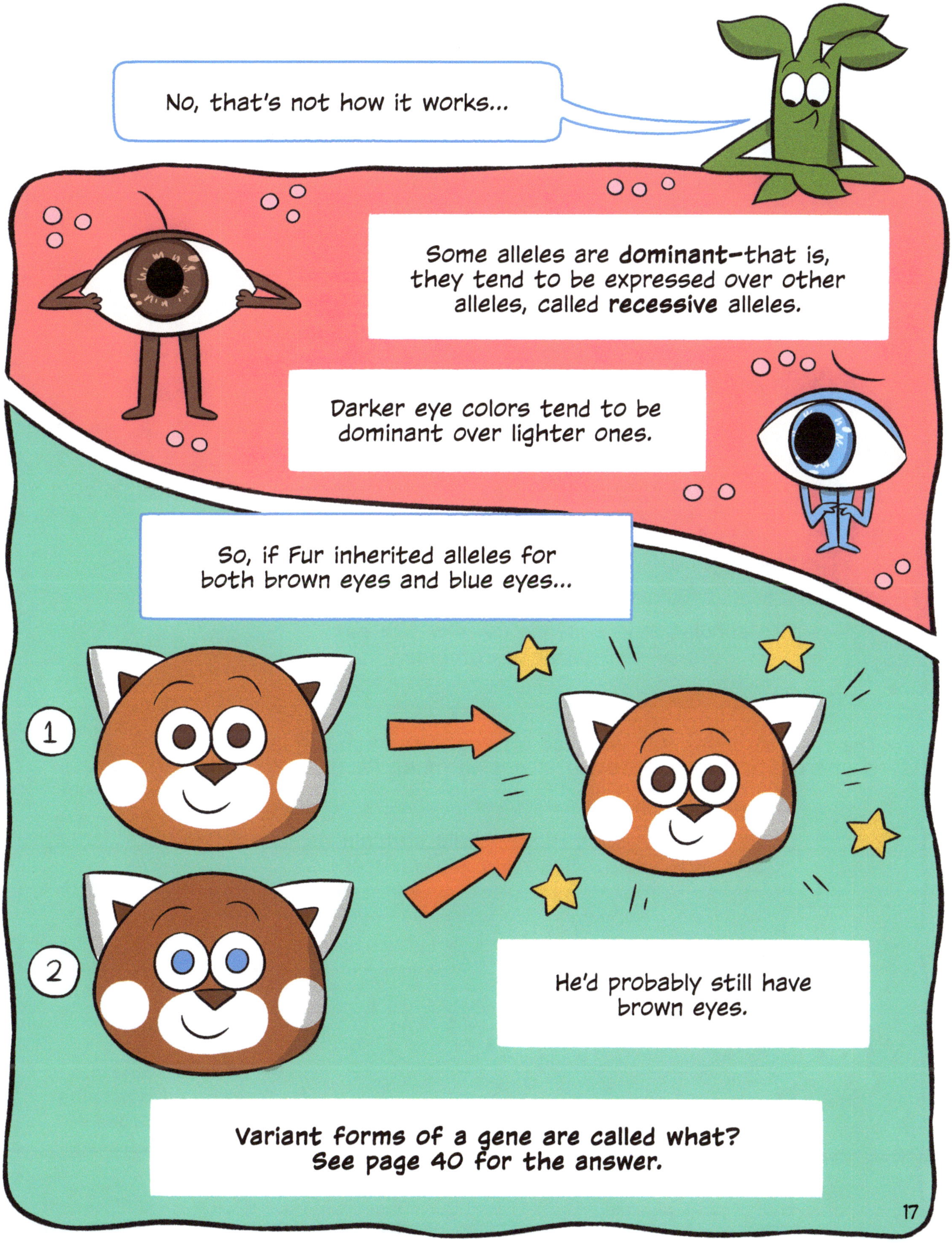
No, that's not how it works...
Some alleles are **dominant**—that is, they tend to be expressed over other alleles, called **recessive** alleles.
Darker eye colors tend to be dominant over lighter ones.
So, if Fur inherited alleles for both brown eyes and blue eyes...
1
2
He'd probably still have brown eyes.
Variant forms of a gene are called what? See page 40 for the answer.

INHERITING GENES
How exactly do parents pass their genes on to their children?
To explain that one, we'll need the simulator.
KLICK!
SIMULATOR
Whoa! Where are we?
Humans and many other living things have special cells called sex cells...
Egg
Sperm
The simulator is showing us how the sex cells of two parents combine.
The mother's sex cell is called an egg. The father's is called a sperm. To make offspring, a sperm must fertilize an egg.
Remember, most human cells contain 46 chromosomes... Sex cells are different...
Egg
Chromosomes
I've counted them... Each sex cell contains 23 chromosomes–only half!
Sperm

Fertilized egg
Chromosomes
During fertilization, the chromosomes from sperm and egg combine... The fertilized egg has 46 chromosomes...
That's 23 from the mother...
...and 23 from the father!
Fertilized egg cell
4 Cells
2 Cells
16 Cells
Fetus
Baby
The fertilized egg cell is the start of a new human. Each time it divides to form new cells...
The new cells contain the same 46 chromosomes... half from the mother...
...and half from the father!

Why are some babies born male and others female?
Well, it all depends on one pair of chromosomes, called the **sex chromosomes.**
There are two kinds of sex chromosome in humans, called X and Y.
X Chromosome
Y Chromosome
Sperm cells contain either an X or a Y chromosome... Egg cells contain only X chromosomes.
XX
XY
In general, babies born with two X chromosomes tend to be female, and those born with an X and a Y tend to be male.

So, if the sperm that fertilizes the egg has an X chromosome, the baby will be female...
And if it has a Y chromosome, the baby will be male!
So, we can find out if the intruder was male or female...
...by seeing if these cells have a Y chromosome or not.
Look! That one's much smaller. It must be a Y chromosome.
So, we're looking for a male intruder!
How many chromosomes does a fertilized egg have? How many are sex chromosomes? See page 40 for answers.

SCIENCE FUN WITH FUR!

PASSING ON GENES

YOU WILL NEED:

- seven plastic cups
- 24 colored marbles (or paper scraps or other counters) 3 blue, 3 green, 3 red, 3 yellow, 3 brown, 3 purple, 3 black, 3 white

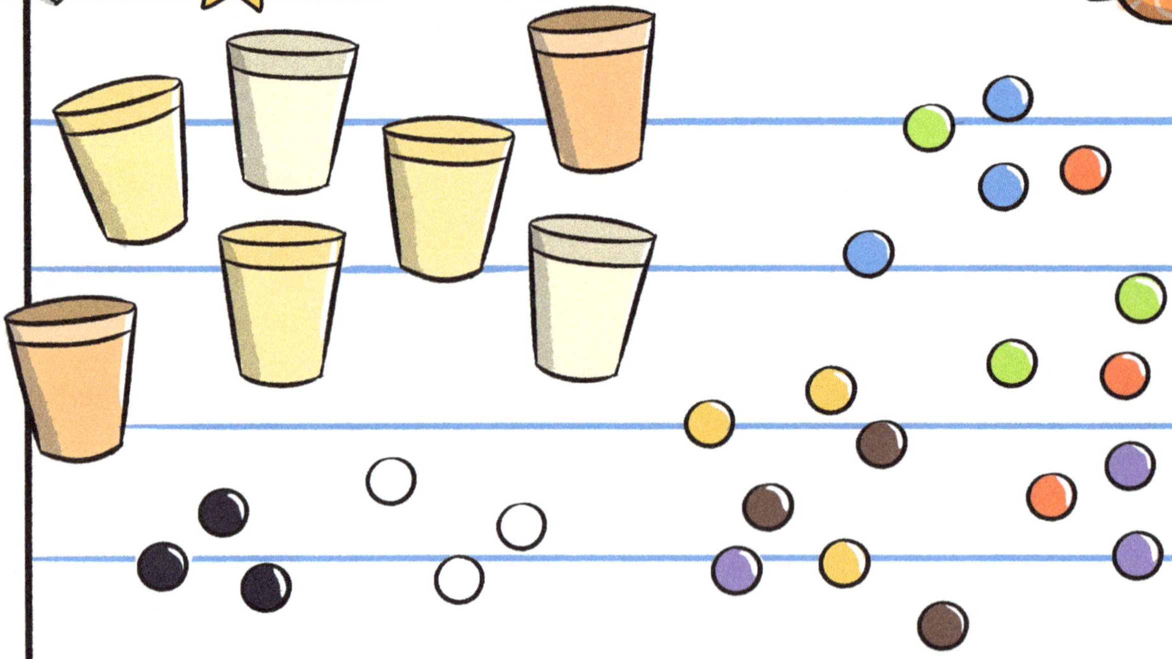

①

Label the seven cups as follows: FATHER'S FATHER; FATHER'S MOTHER; FATHER; MOTHER'S FATHER; MOTHER'S MOTHER; MOTHER; ME.

②

Place six marbles in each cup, as follows:

FATHER'S FATHER: 3 blue, 3 green

FATHER'S MOTHER: 3 red, 3 yellow

MOTHER'S FATHER: 3 brown, 3 purple

MOTHER'S MOTHER: 3 black, 3 white

The marbles represent each grandparent's genes.

(3) Close your eyes. Pick three marbles from FATHER'S FATHER cup and place them in the cup marked FATHER.

(4) Now take three from the cup marked FATHER'S MOTHER (without peeking) and add them to the FATHER cup. These represent the genes your father inherited from his parents.

(5) Repeat the above steps, filling the MOTHER cup with three marbles selected from each of her parents.

6 Now pick 3 marbles from MOTHER cup and 3 from the FATHER. Place them in the cup marked ME. These marbles represent the genes you inherited from your parents.
FATHER
ME
MOTHER
Which marbles ended up in your cup? Reset the marbles, and try the activity again. Did you end up with the same combination of genes?
We inherit the same number of genes from each parent, but the assortment of genes we receive is somewhat random.
This activity uses just six marbles to represent an individual's genes. A human being has nearly 20,000 genes...
ME
Whoa, that's a lot of different combinations!

MUTATIONS
Back in the Nanoship...
Genes remind me of computer code...
They tell the body what to do, much as code tells the computer what to do.
ERROR!
TAP
TAP
TAP
But code sometimes contains errors...
Do genes ever get it wrong?
Indeed! The body's cells divide to make more cells for growth and repair...
But before a cell can divide, it must copy its DNA—so that each new cell has a complete set of chromosomes.

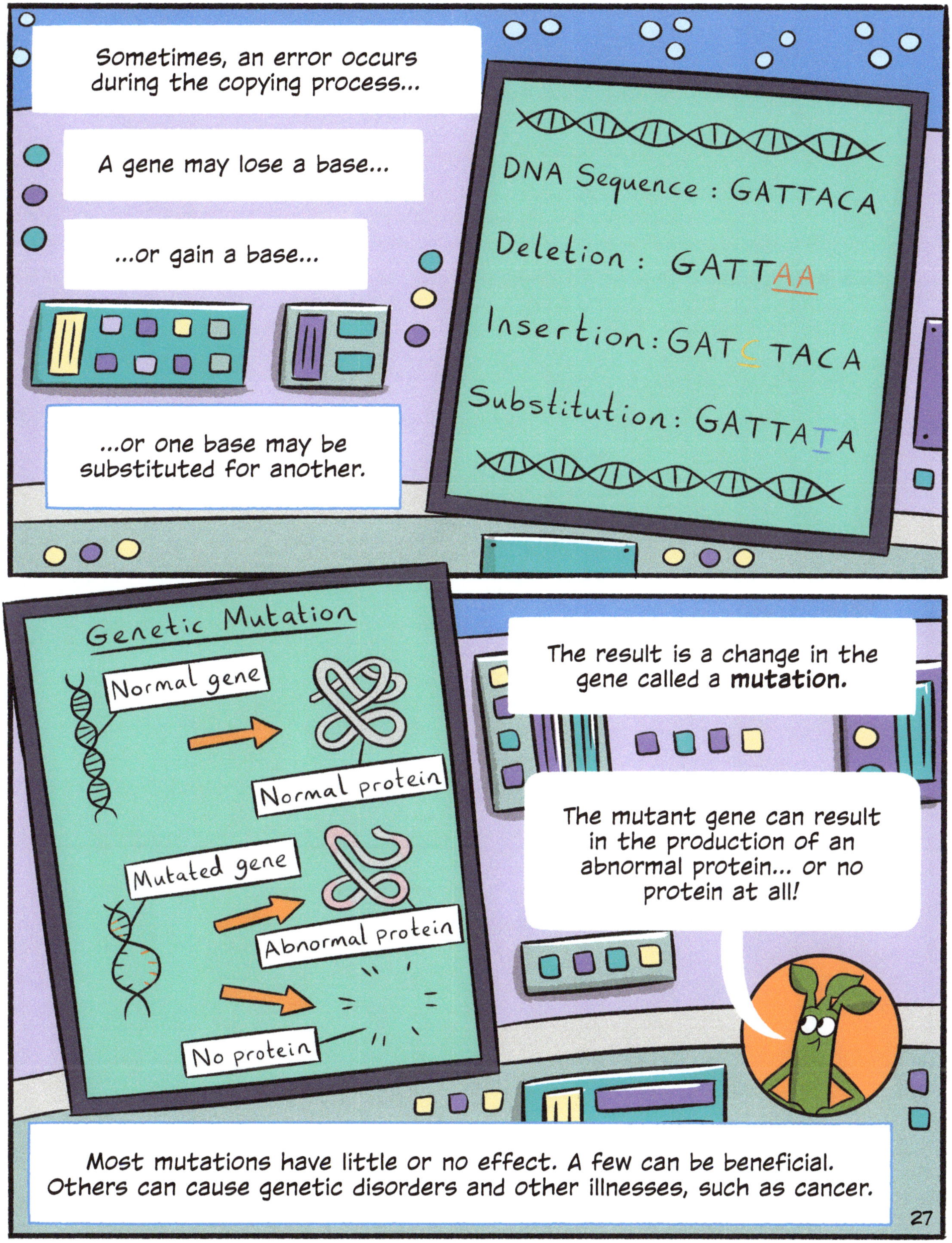
Sometimes, an error occurs during the copying process...
A gene may lose a base...
...or gain a base...
...or one base may be substituted for another.
DNA Sequence : GATTACA
Deletion : GATTAA
Insertion : GATCTACA
Substitution : GATTATA
Genetic Mutation
Normal gene
Normal protein
Mutated gene
Abnormal protein
No protein
The result is a change in the gene called a **mutation.**
The mutant gene can result in the production of an abnormal protein... or no protein at all!
Most mutations have little or no effect. A few can be beneficial. Others can cause genetic disorders and other illnesses, such as cancer.

GENETIC DISORDERS
Some mutations cause disorders that can be passed from one generation to the next...
Examples of such genetic disorders include cystic fibrosis and sickle cell anemia.
Sickle cell anemia is a blood disease that can make a person feel weak or tired, among other serious symptoms. Cystic fibrosis can cause a dangerous buildup of mucus in the body.
Some genetic disorders are dominant...
BEEP
BOOP
Dominant Disorder
Affected
Unaffected
Affected father
Unaffected mother
If one parent has the disorder, roughly half the children will inherit it...

Recessive Disorder
Affected
Unaffected
Carrier
Carrier father
Carrier Mother
Other disorders are recessive–requiring two copies of the mutant gene...
Parents can carry one copy of the recessive gene and not even know it.
It's sad that our genes can cause us problems.
Can we do anything about genetic disorders?
There is something...
Doctors can treat some disorders using **gene therapy.**
For example, they can insert a healthy copy of a gene to replace a nonfunctioning one.
Neat!

IDENTIFYING THE INTRUDER

The intruder has different alleles for eye color...
One of his parents has green eyes. The other has blue eyes.
Eye color
Skin color
Hair color
Nose size
Nose shape
Eye shape
Hair texture
Green is dominant over blue. So, he must have green eyes.
Both parents passed on alleles for red hair. So, he must have red hair!

Now we know what our intruder looks like!
SHRINK
Let's get back to regular size!
VOOP!
Hey! He's stealing my candy!
I just saw the door was open. It's my first time in here. Honest!

I'm not sure that's true!
GULP
FUR'S CANDY JAR
I promise I'll never steal anything again.
My grandpa was a thief and now he's in prison.
FUR'S CANDY JAR
That's one trait you definitely wouldn't want to inherit!

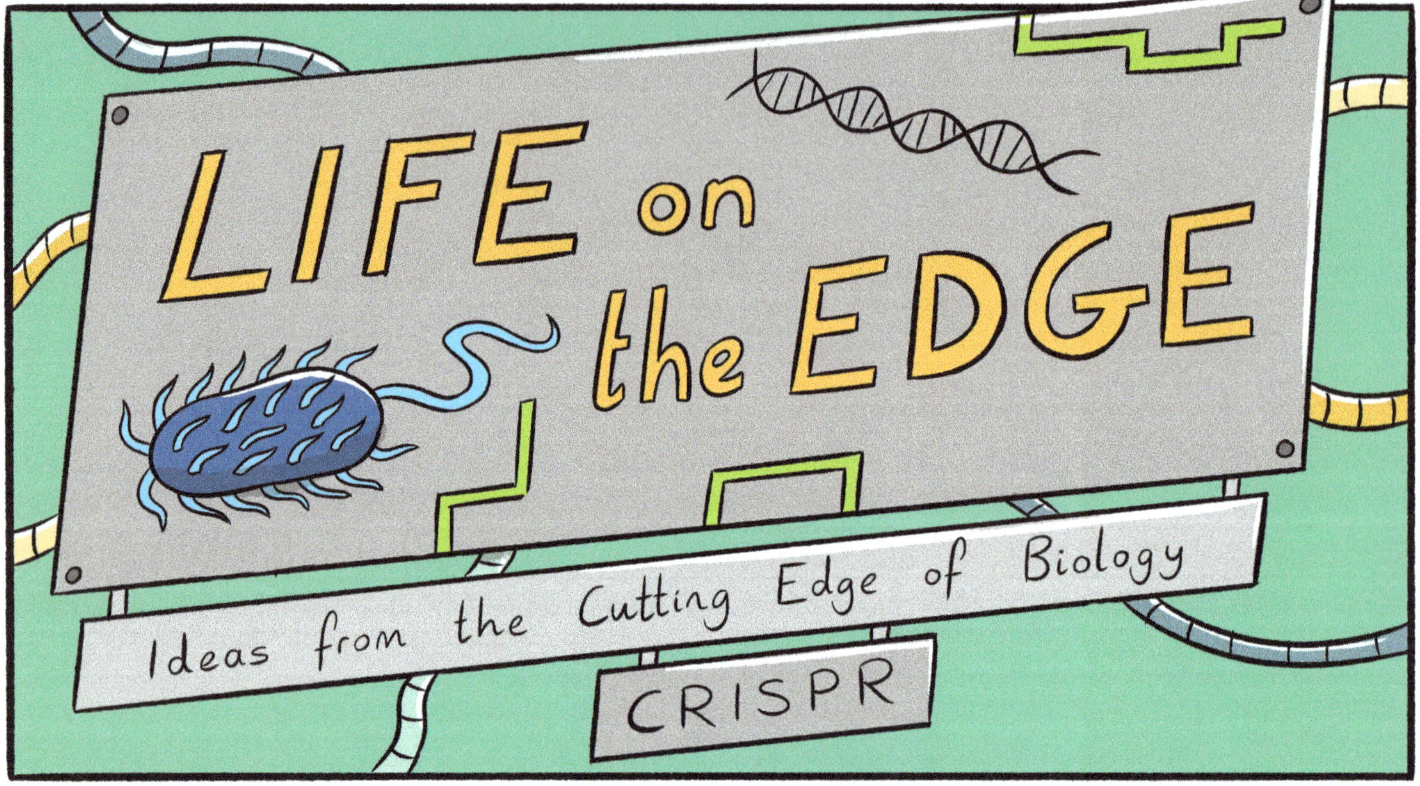
LIFE on the EDGE
Ideas from the Cutting Edge of Biology
CRISPR

The following Wednesday...
Hey, Root, what are you doing? What's **CRISPR?**
CRISPR
Yeah, sounds delicious. I'm starving!
Actually, CRISPR is an exciting new genetic treatment.
Put on these VR goggles. I'll show you!

In nature, bacteria use CRISPR as part of their defense against infection by viruses.
Virus
Bacterium

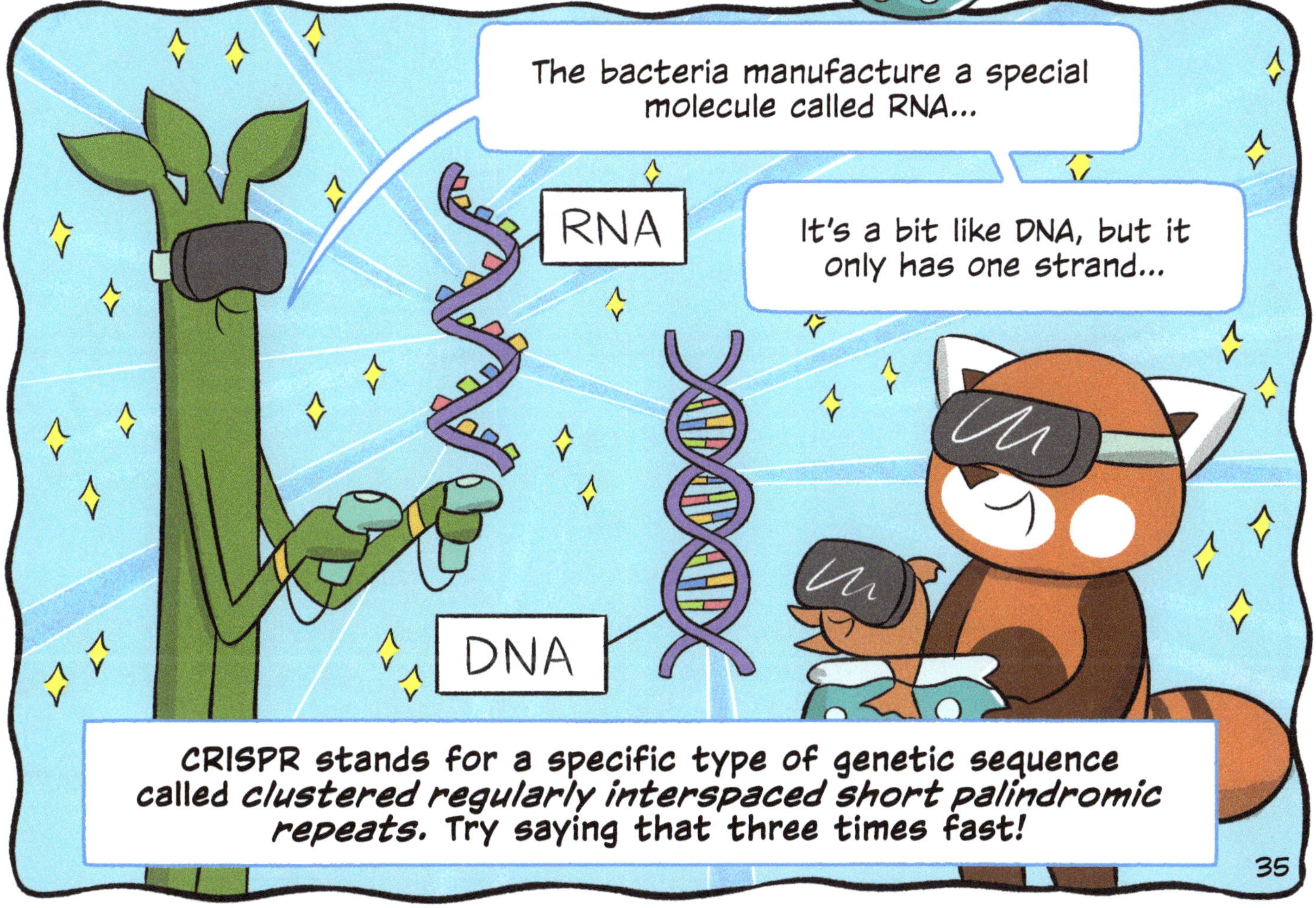
The bacteria manufacture a special molecule called RNA...
It's a bit like DNA, but it only has one strand...
RNA
DNA
CRISPR stands for a specific type of genetic sequence called *clustered regularly interspaced short palindromic repeats*. Try saying that three times fast!

The RNA strand carries a special molecule called an enzyme—kind of like a tiny pair of scissors...
snip snip
The RNA guides the enzyme to the invading virus's DNA, and the scissors cut the DNA to bits!
Hooray for the bacteria!
But how does that help us?
The same system can be used to alter human DNA in medical treatments...

Scientists can use CRISPR to cut human DNA at precise locations...
Snipping out a faulty gene, for example.
Scientists then use the cell's own repair tools to replace the removed gene with a healthy gene...
...helping to cure such diseases as sickle cell anemia and cystic fibrosis!
BEEP BEEP
Neat!

SHOW WHAT YOU KNOW

1. Fill in the blanks.

A. Living things inherit certain characteristics called ______ from their parents. These are encoded in sequences of DNA called ______ .

B. DNA is found in the ______ of the cell. It is stored in threadlike structures called ______ .

2. Does each of the following describe dominant alleles, recessive alleles, or both?

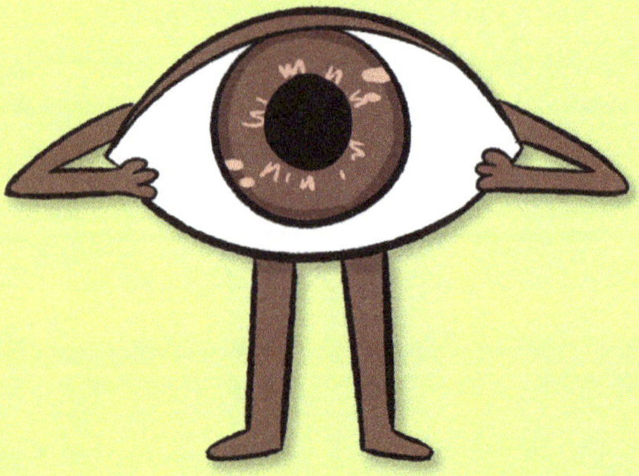

A. variant forms of a particular gene

B. tend to be expressed when only one copy is present

C. tend to be expressed only when two copies are present

D. dark eye color is an example

3. True or false.
A. Human sperm cells can contain an X or a Y chromosome.
B. People with two X chromosomes tend to be born male.
C. Human beings have 23 sex chromosomes.
D. Sperm and egg combine in a process called mutation.

4. Each of these genetic sequences has undergone a mutation. Is it an insertion, deletion, or substitution?
A. AATGC → AAGGC
B. TGCAG → TGACAG
C. GATAC → GAAC
D. CATTA → GATTA
See page 40 for answers.

ANSWERS

page 7: leaves, colored fur patches, fin shape

page 11: chromosomes; nucleus

page 17: alleles

page 21: 46; 2

SHOW WHAT YOU KNOW ANSWERS
pages 38-39:

1. A. trait; gene
   B. nucleus; chromosome
2. A. both
   B. dominant alleles
   C. recessive alleles
   D. dominant alleles
3. A. true
   B. false
   C. false
   D. false
4. A. substitution
   B. insertion
   C. deletion
   D. substitution

WORDS TO KNOW

alleles gene forms that vary from one person to the next.

chromosome a threadlike structure in cells that stores genetic material.

CRISPR a gene-editing technology developed from bacteria.

DNA the ladderlike molecule that holds the genetic code.

dominant describes an allele that tends to be expressed over recessive alleles.

gene a DNA sequence that encodes the instructions for making a particular protein.

gene therapy the use of gene-altering techniques to treat illness.

mutation a change in the sequence of DNA.

nucleus the center of many living cells; contains genetic material.

protein the "building blocks" of living things.

recessive describes an allele that only tends to be expressed in the absence of a dominant allele.

sex chromosomes a pair of chromosomes that help determine biological sex.

trait a particular characteristic of a living thing.

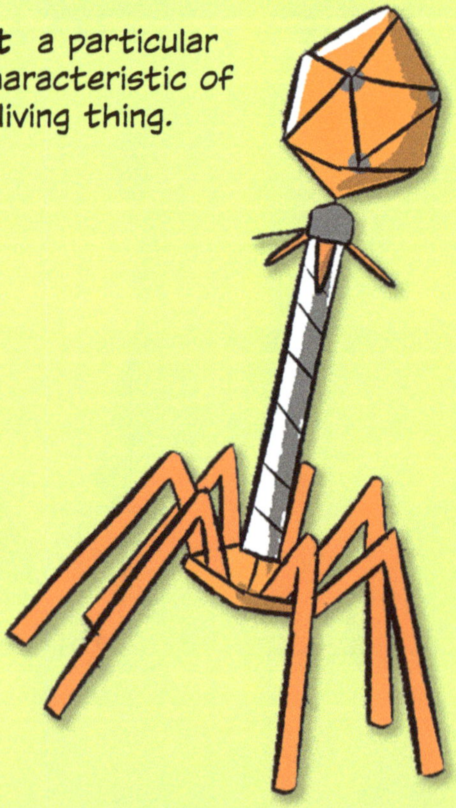

www.ingramcontent.com/pod-product-compliance
Lightning Source LLC
LaVergne TN
LVHW070523061126
851857LV00023B/2133
9780716667551